RAGNAR LOTHBROK

Copyright © 2021
University Press
All Rights Reserved

Table of Contents

Introduction
Chapter 1: A Viking Tale Set in Place and Time
Chapter 2: A Brief Look at Norse Mythology
Chapter 3: Lagertha and Thora Tow-Hart
Chapter 4: Aslaug
Chapter 5: The Sack of Paris
Chapter 6: War and Plunder
Chapter 7: Ragnar and His Sons
Chapter 8: Plundering the British Isles
Chapter 9: Ragnar's Death
Chapter 10: Ragnar's Sons and The Great Heathen Army
Chapter 11: Ragnar Lothbrok: Real or Fictional
Chapter 12: Ragnar Lothbrok: Hero or Villain?
Chapter 13: Ragnar's Story: Early Literary Works
Conclusion

Introduction

From the misty depths of the Early Middle Ages comes the story of a mighty Viking warrior and king named Ragnar Lothbrok. The rich oral tradition of Norse legend, written down hundreds of years after the events happened, sifted the story of Ragnar's exploits through time and poetry to create for Ragnar a lasting image of bravery in the face of the unknown.

The dramatic tale of Ragnar brings up interesting questions. For example, was Ragnar a real person? If so, have the facts about him been exaggerated to the point that they no longer have much basis in reality? And, if its storytellers have reshaped the story, what does this say about what the people wanted to hear? Also, how can Ragnar be called a hero when his most notable feat was invading and plundering Paris? Also, why was Ragnar called Lothbrok, the Old Norse word for Shaggy-Breeches?

Through his legend, one can see a man who had three strong wives and many vengeful sons.

What kind of father and husband could this man have been? Ragnar was certainly no peacemaker. He was tall and powerfully-built, but what about his moral character? Modern people might see this "Scourge of England and France" as a brute and a warmonger, and rightly so – not exactly in keeping with modern mores.

Yet, even modern people can appreciate what he did for the world. We can thank Ragnar's kind for creating trade routes throughout Scandinavia, Britain, Europe, and even into Russia. In expanding his empire, he aided in the settlement of the regions he conquered.

Learning about Ragnar's life, times, and character can give one a new way to view the world. The fundamental struggle of Ragnar's life was between good and evil, but these two concepts are not as clear-cut as one might think. What Medieval Norsemen saw as good, contemporary people might see as wicked, aggressive, and based on self-interest alone. Moreover, what they saw as evil, contemporary people might see as merely the natural way of life on this planet.

Let us begin by delving into the poetry that tells Ragnar Lothbrok's story so beautifully.

Chapter 1

A Viking Tale Set in Place and Time

Like all good stories and all known historical events, the Ragnar legend takes place in a particular time and place. These two factors help explain the tale and draw for the listener or reader a clearer understanding of why things happened the way they did and what challenges to characters faced.

Ragnar's legend is set near the beginning of the Viking Age. The Vikings had made a few conquests before Ragnar sacked Paris, but the era did not get into full swing until afterward. Specifically, the time frame of the story is between 820 and 879 CE.

During the Viking Age, the present countries of Scandinavia and Europe did not exist. Instead, there were various small kingdoms and tribes. Individual kingdoms attacked other kingdoms often, either because of disagreements, to steal plunder and slaves, or to gain a payment for promising to attack someone else instead. The Viking people identified with their immediate tribe and probably did not even call themselves Vikings.

While the name "Viking" means something like "pirate" or "raider," the Norse people of Scandinavia were predominantly farmers and herders. Most farms had just enough land to support the family that lived on them. Both in Scandinavia and in the regions they conquered, the Vikings built their houses of wood and other perishable materials, with only one door and a thatched roof, evidence of which has been found in archeological excavations like Wood Quay.

Some Vikings went on expeditions to raid other countries, but these pirates made up a much smaller percentage of the population than most modern people assume. Often the ones who invaded other countries did so because they did not have enough farmland to support their people. The Vikings followed pagan religions and may have attacked to fight back against the spread of Christianity.

Another reason for Viking conquests may be that they started as trade excursions. When the Vikings arrived to do trade, they were often turned away, and the situation became hostile. For example, British merchants would not do business with those they considered heathens or infidels. This included both the Norse people and the Muslims. Yet, many other Viking expeditions were peaceful trading missions that concluded successfully for both parties.

Personal names of Vikings were often composed of a first name and a nickname. Ragnar's nickname was Lothbrok, an Old Norse word meaning "Shaggy-Breeches," a name he got while trying to win the hand of Thora Town-Hart, as the story goes. A surname often included the father's name. Ragnar is sometimes called Ragnar Sigurdsson, and Ragnar's sons are believed to have had the surname Ragnarsson.

As Vikings traveled, either trading peacefully or raiding, they settled alongside people of other regions. In fact, the people who were raided in the sack of Paris included many Vikings who had opted to give allegiance to the Frankish kings, whose kingdoms later became Normandy in honor of the Northmen who lived there.

Women in Scandinavia had many rights that were not afforded to them in other parts of Europe and Britain. The only exception was women, men, and children who were taken as slaves during raids. While there probably were shield maidens who fought alongside Viking men, most women worked on the farms and kept the home running smoothly.

Another misconception about Vikings is that they were total barbarians. They have been depicted as uncivilized and brutish thugs in some works of fiction. However, Vikings did have a code of ethics. They were also neat and clean, a fact borne out by the wide variety of grooming implements fashioned in a Norse style and found along with Viking remains in an archeological dig in England. No archeological site shows any evidence that the Vikings wore horned helmets, so that is probably an invention of storytellers.

One part of the Viking image is certainly accurate. The Vikings were phenomenal seafarers for their day. They traveled by longboat and had advanced technology for finding their way in unfamiliar waters. For example, they used sunstones to find their direction. Sunstones were made of crystalline rock. When the day was cloudy, they could hold up the rock and what little sunlight there was would light up the crystal to show them its direction.

One Viking custom that may figure into Ragnar's story was that of adoption. Viking families often adopted children in order to preserve their family line. Because parents often died, either in battle or because of the harsh conditions under which the Vikings lived, adoption was also common to give homes to orphans.

The Viking Age in Scandinavia was unique in many respects, having its own customs and beliefs. Life was not easy for the Vikings, but they met the obstacles with bravery. Such was the world in which Ragnar's story takes place.

Chapter 2

A Brief Look at Norse Mythology

Learning about Norse mythology can give one a glimpse of the motives behind Viking raids such as those conducted by Ragnar. The Norse people believed in and honored the Norse gods and goddesses in wars, raids, and even in their daily lives. The Christian counties they sacked or conquered dismissed these beliefs, claiming that they were false heathen religions. However, Norse mythology was considered true by most Scandinavians at that time.

Creation Story

Ginnungagap was a space where the worlds of fire and ice came near each other. At that juncture, frost formed and turned into a giant named Ymir. Audhumla, a mythical cow, also arrived on the scene. The cow licked the ice until she found a man and his three grandsons, one of whom was Odin. The three brothers killed

Ymir and used the giant's body to create the earth, seas, and sky.

One version of the creation myth has Odin and his brothers making the first man and woman out of trees. Another says they popped up out of Ymir's armpits.

Bifrost
A rainbow bridge connected the world of the Aesir with the world of humans. This bridge is called Bifrost and is guarded by the god Heimdall.

Odin
The Norse gods and goddesses were very much a part of the world for people like Ragnar. Viking kings, commanders, and commoners paid homage to Odin with prayers and acts of bravery. Odin was the chief of the Aesir, a group of war gods and sky gods. Although the Norse mythology establishes Thor as the god of war, the Vikings prayed to Odin and honored him with their victories. Odin was, after all, Thor's boss.

Frigg
Chief goddess of the Aesir and wife of Odin, Frigg is the mother of Baldur, Odin's son who was also an Aesir. Frigg might have also been

called Saga, although the experts are not sure of the connection.

Vali
Vali, the god of revenge, was one of Odin's sons and the twin brother of Baldur. Although Vali may not be mentioned in the written records of the Skals stories about Ragnar Lothbrok, Ragnar's sons must have thought of this god while taking their revenge for Ragnar's death.

Thor
Thor was second-in-command among the Aesir - ranking just below Odin. Thor was a war god and is described as the god of thunder. He had a hammer named Mjolnir.

Tyr
Tyr was a mild-mannered sky god and the son of Odin. He was the god of victory. After he was murdered, he went to Hel in her underworld, only returning to the new world after Ragnarok ended and the new world was created.

Loki
Norse mythology recognized that things cannot always go as planned. For example, Loki was a trickster god who made trouble for other gods as well as humans. However, he helped them just

often enough that they were willing to trust him in a pinch. This, of course, made them even more vulnerable to Loki's mischief.

Freyr and Fryja

Freyr was one of the Vanir, gods who represented the earth, prosperity, and fertility. The Vanir often fought against the Aesir. Freyr was the god of agriculture and male fertility and sexuality. He figures prominently in the great battle of the gods and giants during Ragnarok. Fryja was Freyr's twin sister, who was the goddess of female fertility and sexuality.

Njord

Njord was the patron god of the sea and sailing and one of the Vanir.

Death and Hel

Honor was paramount for the Vikings. If another tribe impugned their honor, they would strike back with bloody attacks. Even the Norse mythology points to the importance of tribal loyalty and honor. Those who died in battle were given a hero's send-off with burial at sea. Anyone who died in battle was destined to go the icy hell of Niflheim, ruled by the Norse goddess Hel. Hel was the daughter of Loki, the trickster god, and gained her place in the underworld when Odin sent her away from Asgard, the home of the gods.

Valhalla

Odin rules Valhalla, where those who have died in battle go after their deaths. They have to be worthy of Odin's respect, whether they died in battle or some other way. In this respect, Ragnar might go to Valhalla if he died at the hands of King Aella. But, if he died of the plague, he would go to Hel's underworld instead.

The dead in Valhalla are continually fighting. As they fight, they improve their fighting skills to prepare for Ragnarok. When one battle is over, they sit down at the table in the great hall of Valhalla for a celebration and then go back to fighting again. According to Old Norse texts, Valhalla is located within the underworld, either along with or as a part of Helheim.

Ragnarok

In Norse mythology, Ragnarok referred to the end of the world. Sometime in the future, when Ragnarok took place, the world would suffer a devastating and long winter that would be three years long and with no summers. Fighting would erupt throughout the Norse world. Family members would attack each other. Then, earthquakes and other cataclysmic events would take place. The earth would be plunged into darkness and crack open at the tectonic plates. Next, the gods and giants would begin a long battle.

As Ragnarok built to a close, the gods would fight with a wolf and a serpent. The serpent's venom would kill Thor. The wolf would swallow Odin. The world would be engulfed in fire and then drowned in a flood.

After Ragnarok, the earth would rise out of the sea and be green and beautiful once more. The sons of the gods would return. Only two humans would survive, and they would repopulate the world. Unlike in the Christian end-of-world story, the world formed would contain both kind people and dangerous foes.

As one reads about Ragnar Lothbrok's legend, one can see how many of these gods, people, and animals were incorporated into the stories handed down by the Norse storytellers, the Skalds.

The mythology was more than a collection of exciting stories for the Norsemen of that time. They chose how to live based on their views of what the gods did. They celebrated their victories in honor of the god Odin. And, they incorporated information about the gods into their oral histories.

Chapter 3

Lagertha and Thora Town-Hart

Danish King Sigurd Hring and Alfhild, the daughter of King Alf of Alfheimer, were the parents of Ragnar Lothbrok. After giving birth to Ragnar, Alfhild died. Ragnar eventually became king of Denmark and Sweden. However, he lost control of Sweden after putting a sub-king in his place there.

Ragnar has sometimes been portrayed in stories and films as a lowly farmer who became king after many successful raids. However, other sources point to his royal lineage to assert that he was a nobleman who became a great warrior. He was physically strong, but also intelligent and generous with those that fought under him. He had many warships and troops to command.

Ragnar first married Lagertha, a shieldmaiden who was renowned for her fighting skills. Ragnar was impressed by the stories about her and went to seek her hand. When he arrived, he had to kill a bear and a hound that she had set out to guard her home against him.

Ragnar and Lagertha were married and had two daughters and a son named Fridlief. But, Ragnar was always angry with Lagertha because she had put the bear and hound out to stop him from arriving. He divorced her when he met Thora Town-Hart, but when he needed help, Lagertha sent him 120 of her ships.

Some historians believe that Lagertha was not an actual woman. They say the mentions of Lagertha in the sagas known to exist may refer to Lathgertha. This may be another name for Thorgerd, a Norse guardian goddess who may also be associated with the goddess Freyja.

Ragnar acquired the name Lothbrok after killing a serpent in a bower in the lands of Earl Herruth of Gautland, in what is now Sweden. Earl Herruth had given his daughter a small heather snake to raise, but when it grew extremely large and annoying, he proclaimed that anyone who killed the serpent would get the pile of gold that was hidden beneath it as well as the hand of his daughter Thora.

Ragnar cunningly made a suit of fur clothing and dipped it in tar. Then, he went to the bower, took out his spear, and attacked the snake, sinking his spearhead deep into the serpent's spine. Wearing the fur outfit, he was protected from the serpent's venom, which the serpent released on him as it died. Because Lothbrok is the Old Norse word for shaggy breeches, Ragnar is also known by the name Ragnar Shaggy Breeches.

However, the name Shaggy Breeches is not mentioned until 12th-century writings about Ragnar. Some critics of the Ragnar Lothbrok legend have a different explanation for the name. Because he suffered from diarrhea while at the court of Horik, his clothes might have been soiled with feces. Those who noticed the dirty appearance of his clothing might have made fun of him. His pants may have looked like they had been dipped in tar. Thus, the Shaggy-Breeches moniker may have been more of a joke than a tribute to Ragnar's foresight.

Ragnar and Thora were married at a large feast. Afterward, the couple returned to Ragnar's kingdom. Ragnar and Thora had two sons, named Eirek and Agnar. Other accounts list the sons as Radbard, Dunwat, and brothers Siward, Bjorn, Agnar and Iwar. Later, Thora became ill.

When she died from the illness, Ragnar was heartbroken. So, he handed over his kingdom to the care of sub-kings and went back to raiding other kingdoms.

Chapter 4

Aslaug

According to legendary accounts, Ragnar fell in love with Esbern's daughter and vowed to have her. So, instead of going directly to the daughter, he courted the father, showing him every kindness and attention. The father had the girl guarded, as he understood what Ragnar wanted. Ragnar dressed like a woman to get close to the girl. He seduced her, and she became pregnant, later giving birth to Ubbe. Ragnar and Esbern's daughter never married.

The Saga of Ragnar tells the story of another wife named Aslaug. A king named Heimir heard of the death of Aslaug's parents Brynhild and Sigurd. As she was one of his foster children, he was concerned for her. He worried that she would be killed or taken, putting an end to their line. So, he put his foster daughter in a harp to travel away from danger. He also stocked the harp with has luxurious clothing and as many jewels and as much gold as it would hold.

During the trip, Heimir dismantled the harp at various remote stops to let her out for a while each time. He fed her with a unique herb that sustained her and even helped her grow. Heimir stopped at the farm home of Grima and Aki. Aki, the old man, asked Heimir what kind of person he was, and he told her he was a poor beggar. Grima saw the exquisite clothing sticking out from the harp and a costly ring sticking out from Heimir's rags and surmised that the "beggar" was actually wealthy. Guessing that they could take his expensive things, she decided they should kill him.

Aki did not want to do it, but after Grima pestered him for a while, they did kill him. Heimir's death cries were so violent that the house pillars fell down and an earthquake shook the ground. When Aki and Grima found Aslaug, they took the daughter for their own, calling her Kraka, meaning crow. Since the girl was so beautiful and the couple was so ugly, they decided the only way to pass her off as their own was to shave her head, dirty her face and give her rags to wear. They set her to work doing the most strenuous and disagreeable tasks on the farm.

Later, Ragnar heard of her beauty and wanted her to be his wife. He had been sailing near Kraka's home and sent his men to see if she really was beautiful. They returned and said she was and brought her back to see him. He offered her a shirt that had belonged to Thora, but she refused it and went back home. However, the next time he sailed in the area, he stopped again. This time, Kraka told Grima and Aki that she knew they had killed her father. She wanted to leave, but before she got away, Grima cursed her, saying that something terrible would happen if they did not wait three days after their wedding before consummating their marriage.

Kraka told Ragnar about the curse, but he did not want to wait. They were married at a great feast. Before the three days were up, Ragnar insisted they have sex. Kraka said that if they did, the child they conceived would be born without bones. She did have a baby, and it did have no bones. They named him Ivar, and he became known as Ivar the Boneless. However, his cartilage grew in the place of his absent bones and became strong. Ivar could not walk, but he had himself carried along into battle to advise his brothers. In this story, Bjorn, Hvitserk, and Rognvald were also born to Ragnar and Kraka.

Ragnar met and married another woman on one of his journeys. Ingibjorg, daughter of King Eystein, was the most beautiful woman in his kingdom. Ragnar wanted her for a wife, and his friend, King Eystein, promised her to him. When Ragnar returned home, Kraka told him that she was not the daughter of Grima and Aki but was the daughter of King Sigurd. Her real name was Aslaug.

Aslaug was pregnant. She told Ragnar the child would be born with an eye that looked like it contained a snake. If she right about this, she said, he must stay. If not, he could go to Ingibjorg. However, the child was born just as she said. The child became known as Sigurd Snake-in-the-Eye or Sigurd Worm-in-Eye.

Chapter 5

The Sack of Paris

Charles the Bald of Francia (what is now France) gave Ragnar Lothbrok some land in Frisia in 841. However, Lothbrok lost the land as well as the king's friendship. Vikings had begun to raid Paris in 799, and raids had continued there off and on until, in 845 CE, Ragnar set out to make his mark there. In March, he arrived with 120 ships and 5000 troops. Ragnar commanded the ships as they entered the Seine and invaded Rouen, which later became the capital of Normandy. Then, Charles the Bald got together an army and split it with one side defending each bank of the Seine.

When Ragnar defeated one of the divisions of Charles the Bald's Frankish army, he captured 111 prisoners. To honor Odin and to cause fear among the Frankish troops, Ragnar had all 111 of them hanged. The rest of Charles the Bald's army retreated immediately, leaving Paris to Ragnar and his men.

When Ragnar's Vikings arrived in Paris on Easter, they raided and plundered the city. While they were camped there, a plague broke out. Ragnar's men had destroyed church property as they plundered Paris. Perhaps feeling guilty and concerned about how to handle the crisis, Ragnar's men talked to one of the Christians they had just captured in Paris about how to get the help of their Christian god. So, they said a prayer to the Norse gods and fasted to appease the Christian one. The plague was soon over. After the battle, Ragnar said that taking Paris was easy, except for the plague that, according to him, came from the deceased Saint Germain of Paris.

Since Charles was facing regional revolts and distrusted his brothers, he was not able to mount an adequate defense. He needed time to settle the disputes within his own kingdom before he would be ready to face the Vikings again. His only option was to pay off Ragnar so he would leave Paris. The ransom he paid was 7,000 livres of silver and gold. Paris paid the same amount 13 more times to prevent attacks.

Ragnar, having received the first payment of 7,000 French pounds, took his ships and men and sailed back down the Seine towards home. On his way out of Frankish territory, Ragnar continued to pillage along the coast.

The story of the sacking of Paris is told in Old Norse sources, but it is also a well-known fact to historians. Records do remain of this battle, taken down at the time it happened. While the historical records do not use the name Ragnar specifically, the conqueror was listed as a Viking named Reginherus. With the similarity of the names and the translation into another language, Reginherus could easily have been Ragnar.

According to some sources, Ragnar was wounded during the Sack of Paris, and these injuries eventually caused his death. However, most sources clearly point to another kind of death for Ragnar.

Chapter 6

War and Plunder

Before he died, Ragnar Lothbrok is said to have attacked many villages and cities, usually when they were poorly defended or vulnerable in some other way. These battles were a way of life for people like Ragnar during the Viking Age, and they looked forward to the fights with eager anticipation.

Sometimes the Viking warriors would go overland, but more often, they would go by ship or longboat. The typical raid began with a surprise attack on unsuspecting townspeople. The Vikings would then take their plunder. While they were at it, they would seize people, some whom they sold as slaves. They also kidnapped more noteworthy people and demanded money as a ransom so they could cash in even more on their adventure. They often attacked churches and monasteries first, as these were the places where the people they were raiding were most vulnerable.

Before Ragnar met Lagertha, Ragnar's grandfather Siward, King of Norway, was killed by the King of Sweden. (At least one of the Old Norse texts says that Siward was Ragnar's cousin, and Hring was his father. This seems to fit with other mentions of Ragnar being in the line of "Ring," although Siward could still have been his grandfather rather than his cousin.)

Many women came to see Ragnar to ask for his help. The Swedish king had imprisoned the women in a brothel. One of them was Lagertha. The women came out to do battle against the King of Sweden. Ragnar killed his grandfather's murderer. Three years later, he fought with the Skanians against the Jutlanders and came away victorious.

After Ragnar's son Ubbe was born, he went on an expedition to the Dardanelles, called the Hellespont in Saxo Grammaticus' Gesta Danorum. He took his sons with him to conquer the Hellespontines. The sons of the king of Hellespont were married to daughters of a Russian king when Ragnar killed the

Hellespontines' king. The Russian king, afraid of Ragnar and his troops, fled.

Some sources say that Ragnar was not a king, but was a lord under the Danish King Horik. By this account, Ragnar went on from sacking Paris to sack Hamburg. Hamburg had been placed in charge of the Saxon territory by Louis the Pious. In addition to governing the Saxon territory, Hamburg was to introduce Christianity to the heathens in Scandinavia.

After Ragnar invaded Hamburg, Louis the German sent word of his demand that King Horik pay reparations to East Frank. King Horik made peace with Louis the German and stopped supporting Ragnar and other Vikings in their raids. However, the Vikings continued to attack France until the Siege of Paris in 885, when the Vikings were defeated.

Lothbrok journeyed to various parts of the known world, sacking and plundering along the way. He attacked parts of Russia and Lithuania, and by some accounts, even traveled to China. Vikings often stayed behind these areas, establishing new settlements.

At one point, there was a group of Danes that helped a tyrant named Harald fight against Ragnar in a civil war within his kingdom. Ragnar fought and won back the territory. He not only took captives, but also tortured them to teach others not to go against him.

Ragnar had fought and gained back all his land. Now, he wanted to take England. However, his ships were wrecked there upon arrival. Ragnar and his men made it to shore. After that, they raided and plundered wherever they went.

Chapter 7

Ragnar and His Sons

As written in the Saga of Ragnar's Sons, Ragnar had relationships with his sons that might at first seem unusual in modern times. However, both then and now, such relationships may exist occasionally. The heart of the issue was that Ragnar at times supported the welfare of his sons, and at other times, was extremely competitive with them. He wanted them to be famous and accomplished, but not at the expense of giving up his own glory.

Ragnar's sons Eirek and Agnar spent every summer going out to raid and plunder. After they had gained fame, they wanted to seek even greater triumphs. They requested that Ragnar give them ships and troops so they could go out and conquer more challenging foes. Then, they asked for his advice. They wanted to know where to find the most challenging fight. Ragnar told them about a place where the pagan people used powerful magic to overcome any attackers.

Ivar, who was always ready with helpful advice, told the brothers and their troops that there were two cattle in the garrison that were so ghastly that enemies had always run upon hearing their bellows and seeing their ugly appearance. He told them they would be fine as long as they did not panic and run away.

The older brothers left young Rognvald behind with the ships and went to face their enemies. When the opposing troops sicced the cattle on their troops, Ivar asked for his bow and arrow. He took aim at them and killed them both. Then, Rognvald told the troops in the ships that they should all go out and join in the battle so the older brothers would not get all the glory. Rognvald fought valiantly but was killed. The rest of the brothers and the troops drove the enemy out of their fortress, chased them away, and then returned to the fortress, stole all the treasure, and burned the place to the ground.

After Ragnar married Aslaug, his sons Eirek, Agnar, Ivar and their younger brothers were old enough to do battle, so Ragnar took them out to raid kingdoms and small islands. The sons wanted to be as famous as their father, so they went out and started wars to show their prowess in battle. They took much of his land without his approval.

Eirik told King Eystein that he wanted him to govern Sweden. He also wanted to marry that king's daughter. The king consulted with his chieftains and they decided to defend the kingdom against Eirek and his brothers. The large number of troops the king sent overwhelmed the brothers and their men. Agnar was killed. The king's men captured Eirik. The king offered Eirek peace, along with his daughter and plenty of money. Eirik only wanted to choose the day of his death.

Eirik was killed on the day he had chosen. When Auslag heard of it, she sent Bjorn Ironside, Hvitserk, Sigurd Snake-in-Eye, and Ivar the Boneless to avenge Eirik's death. Ragnar's sons took an army and sailed to Sweden. Meanwhile, Queen Aslaug took 1500 knights with her to fight in the battle.

King Eystein was killed, and Aslaug and Ragnar's sons won the battle. Aslaug went back home, but the brothers went from town to town, continuing until they were bested in a southern kingdom. Ivar the Boneless had a plan, though. He had the brothers and their troops wait in the forest until the town no longer expected them. Then, they raided the town, setting fires there and killing all the children. They took all the townspeople's goods away as they left.

However, Ragnar was not pleased that his sons and wife did not include him or ask him for advice. Ragnar told Aslaug he would gain back prominence among his clan.

When Ragnar's sons came to the town of Luna, they talked to an old man, asking him the way to Rome so that they could raid it. But, the man told them it was too far.

Chapter 8

Plundering the British Isles

If Ragnar Lothbrok was a real person, he might have been the leader of a group of Vikings who attacked Ireland. Both historical accounts and the Old Norse literature describe Viking raids that happened in Ireland between 832 and 845 CE. In the Gesta Danorum, Saxo Grammaticus says it was Ragnar Lothbrok who envisioned, commanded the sixty longboats that arrived in the British Isles and led the raids on Ireland. Historical documents name one Turgesius as the mastermind of these 837 CE raids.

Historians believe Turgesius may have gone by another name as well. Some say he was the son of Harald Harfagri, the first king of Norway. However, others say that the timeline does not match up for Turgesius to be Harald's son. Those who put credence in a historical Ragnar say Turgesius was Ragnar Lothbrok.

During and after the attacks on Ireland, Turgesius (or Ragnar) established settlements and ruled them from Dublin. After one of the Viking leaders was killed later, the Vikings left Dublin, leaving behind those Norsemen who had settled there. In 841 CE, the Vikings, presumably led by Ragnar Lothbrok or another Viking like him, returned to Dublin and conquered it. From then until 845, the Vikings raided and plundered territories from Leinster to the Slieve Bloom Mountains. The Irishmen finally defeated the Vikings for a time in 847, until the sons of Ragnar returned in 856 or 857 CE.

Since the Vikings were in the British Isles for several years, they need places to stay during the winters. To this end, the Vikings set up naval encampments, which they called longphorts. The Annals of Ulster describes two such encampments, one in Dublin and one on Ulster Island. Archeologists agree that there were at least two winter camps, and have found Viking artifacts and remains at a site near Dublin Castle.

Most of the evidence for the Viking raids and the conquest of the British Isles does not settle the question of whether it was the legendary

Lothbrok who led them. However, those who choose to believe that the legend was grounded in fact assume that they were. They base their assumption on the Norse literature that came from Skaldic poems. They also cite Medieval Irish and British texts that refer to the leader of the Vikings at that time as Ragnar, Reginherus, Rognvald, or Ragnall. The Norse literary tradition clearly shows a preference for the raider and conqueror of the British Isles during the mid-800's as being led by Ragnar.

According to the legend, Ragnar continued plundering the British Isles, including Britain, Scotland, and Ireland, until a certain King Aella put an end to his conquest.

Chapter 9

Ragnar's Death in England

British historical records describe the events relating to how England was conquered by a Viking that was the father of those listed here as Ragnar's sons. Old Norse texts also tell the story, albeit in perhaps a more literary way. By all accounts, the attack began in the mid-800's CE.

According to literary sources, Ragnar heard about his son's conquests and decided to go out and boost his own fame with some fighting of his own. He ordered the construction of two large ships. He wanted to go to England. Aslaug warned him that he should have made longboats instead, and that he would not fare well in the large ships. Still, she gave him a special shirt to show her gratitude for his gift of Thora's beautiful embroidered shirt. It was made of wool and had no seams. As long as he wore it, any wounds he might get would not bleed.

Ragnar waited for the ice to break and then set sail for England. The wind was so harsh that the ships broke up against the English shore. King Aella, the ruler of England at the time, had sent scouts to find out when Ragnar landed. Meanwhile, the king gathered a considerable fighting force. He told these men that if they found Ragnar, they should bring him in alive to prevent Ragnar's sons from avenging his death.

King Aella's army was much larger than Ragnar's. Ragnar did not have a hope of succeeding. He fought viciously and killed many of Aella's men, but Ragnar's own army was utterly wiped out in a very short time. He was taken prisoner at spear-point.

In the story, King Aella asks Ragnar who he is. When he does not answer, the king says he will suffer more. He commanded that the foreign king (who was Ragnar) be put in a pit of snakes until he would tell them that he was indeed Ragnar. If Ragnar had done that, King Aella would have had him let out of the pit.

As Ragnar Lothbrok sat in the pit of snakes, the vipers did not bite him at all. King Aella's men were puzzled. However, the king ordered them to take away his clothes. When they did, the snakes bit him. The men still did not know whether he was Ragnar or someone else.

An Old Norse text contains the death poem Ragnar composed, memorized and recited before his death. The poem lists and briefly describes battles Ragnar fought during his life. He ends with a phrase that became associated with defying death: Laughing, I shall die. Some proponents of the historicity of the Ragnar legend say that Ragnar would have only composed a few of the stanzas with Ragnar's court Skald composing the rest of it to be sung at his funeral.

Upon reciting the poem, Ragnar died.

One final note on Ragnar's death: a monk at Saint Germain gave a religious explanation. He said the God struck down Ragnar for his offenses against Christianity. In this version of the story, God took vengeance on Ragnar's body, causing it to swell to bursting. Critics have more recently said that this was an outward description of something that happened to Ragnar internally. He may have suffered from severe diarrhea before his death due to being infected with the plague.

Whatever the actual cause of Ragnar's death, the literary versions favor him. And, while the majority of Norse literature only gives a hero's death to those who died directly in battle, these stories lift Ragnar as a hero on his way to Valhalla.

Chapter 10

Ragnar's Sons and The Great Heathen Army

The one point of agreement among historians about the legend of Ragnar is that Ivar the Boneless, Halfdan, and Ubbe, men listed in the story as Ragnar's sons, were the very real Vikings who led the Great Heathen Army as it attacked England in 865 CE. This Great Viking Army consisted of Norse soldiers and was larger than any previous expeditionary force. Their aim was not to attack and plunder, but to conquer. Historians estimate the size of this military force to be about 1,000 men or more, based on the number of soldiers each ship could carry.

The Great Heathen Army, as the English called it, battled from 865 to 878 CE. As they went, they gained control of cities and rural communities along the way. Starting at East Anglia, the army of Ragnar's sons marched to the last region of Wessex, where Alfred the

Great stopped them on January 8, 871 in the Battle of Ashdown.

The Great Vikings got reinforcements from Scandinavia and continued fighting until Alfred defeated them during the Battle of Eddington. At that point, the Vikings signed Alfred's peace treaty (The Treaty of Wedmore), which gave the Vikings control of large parts of northern and eastern England, where many Norsemen settled on farms and worked the land for the rest of their lives.

The Saga of Ragnar Lothbrok continues after the description of the legendary Viking's death. Since King Aella did not know for sure who it was in the snake pit, he wanted to find out. He was still concerned that Ragnar's sons would avenge his death, taking away his kingdom and possibly his life. So, he sent an envoy of soldiers to speak to Ragnar's son Ivar.

Ivar, Sigurd Worm-in-Eye, Hvitserk, and Bjorn Ironsides were relaxing in a drinking hall when King Aella's troops arrived. The soldiers told Ivar of Ragnar's death. The brothers asked what happened, and the troops told them about the battle and Ragnar's imprisonment and death.

All the brothers, of course, were furious – at least all but Ivar. Ragnar's son Hvitserk said the brothers should kill the messengers. However, Ivar said they should not only let them go, but should even give them anything they needed for their journey. Ivar also suggested they should accept reparations for their loss, but the other brothers said they would never accept payment for their father's death.

After King Aella's men left, the brothers discussed how to exact their revenge. Ivar alone disagreed. He blamed the incident on Ragnar himself. The other brothers declared that they would take their revenge, regardless of what Ivar thought about the matter.

Ivar went separately to see King Aella. The king offered to make reparations, and Ivar suggested Aella give him the amount of land that could be covered by an ox-hide. As the story goes, Ivar killed a large bull, softened its hide, and cut it into strips to make one very long thong. He then went out and used the thong to mark off enough territory to make a large town. He had a vast town hall built as well as many houses. The legend says the name of the town was Lundunaborg.

Rumor had it that Ivar was feigning friendship while plotting something devious. So, the brothers sent him goods, which he had requested. The rumors proved to be correct when Ivar gave the goods to the strongest men in the area and stole these warriors from King Aella's command. Ivar sent word to his brothers that they should gather a great army. They followed his advice and created the largest Viking army ever. Presumably, the fact behind this legend was the Great Heathen Army.

However, the brothers would not listen when Ivar advised them to strike Aella's weakened troops immediately. They said they had already decided to do that before they heard from Ivar, so they did not need his advice. Ivar was so angry that he did not send the troops he had gathered to help the brothers. He even went to King Aella and advised him on how to make war on the other brothers.

Ivar was as two-faced as ever in this legendary exchange. When King Aella was seized, Ivar advised that he be treated as harshly as Ragnar had been treated. The brothers then carved an eagle on Aella's back and the king died. The brothers decided their father's death had been avenged. Then, while the other brothers continued to raid and plunder, Ivar ruled over King Aellla's territory for the rest of his life.

This last bit of the tale may have been used to explain the presence of Norsemen in English communities. Recent archeological excavations have unearthed evidence of Nordic men living on the same farms as English women. Whether the Saga of Ragnar Lothbrok is an exact historical account or not, the fact remains that these families did exist in England in the Early Middle Ages.

Chapter 11

Ragnar Lothbrok: Real or Fictional?

According to many historians who have investigated the legend, the character Ragnar Lothbrok does have some historical validity. There was likely a person (or several) behind the legend, even if the name was not exactly the same.

Several historical figures have been suggested as the basis of the Lothbrok character. One possibility is Reginherus, who went down in history as the Viking leader who sacked Paris in the mid-800's. Various kings of Medieval Scandinavia and Ireland have been considered as the basis of the Ragnar legend as well.

Some critics of the legend cite this fact of several names being ascribed to the same character as a reason to believe that no one person could possibly be identified as Ragnar

Lothbrok in all the sources. However, there have been other historical characters who have been known by different names in different countries and regions. This may be partly due to variations in language or dialect. So, this discrepancy alone may not be enough to discount the historicity of the character entirely.

Other historians propose the theory that the Ragnar character is based on an amalgam of several actual men. If this is true, then Ragnar's sons may actually be the sons of several known kings and leaders. Since they fought together in 865, the same year that Ragnar probably died, storytellers may have created a story that tied them together as brothers.

Because the story was handed down in the oral tradition, many people find it hard to accept it as fact. Somehow written words seem to carry more weight. However, the truth is that, for many peoples, the oral tradition is the only contemporary accounting of events and people that exists, and this is almost certainly the case of the Vikings of Ragnar's time. Still, the facts would be easier to discern if there were records available that were written when Ragnar was alive. Such records would be frozen in time and not subject to later distortions by the storytellers.

The cultures of the
Early Middle Ages were rich in creativity and
intellectual accomplishment. Their technology
was advanced, their artists were creative, and
their craftspeople highly skilled. However, they
apparently did not have written histories - at
least not during Ragnar's time. If there were
documents written during his time, they have not
been found or no longer exist.

Later in the Medieval era, in the 1100s, a
Medieval scholar named Snorre Sturlson wrote
down the stories he had gathered. Many of the
stories came down from the 800's, the same
time frame when Ragnar lived. Scholars today
suggest that Ragnar died sometime between
845 and 865 CE, so these texts were written
over 200 years after the events might have taken
place.

What is certain is that the men listed in the saga
as Ragnar's sons did attack England in 865 CE
as leaders of the Great Heathen Army. They
battled in small Saxon communities for 14 years.
This fact of the sons' historical existence and
actions may lend some credence to the legend
of Ragnar being based on one particular man
rather than an amalgam of several notable men.

Alternatively, it could merely be that the storytellers shaped the story to reflect historical events within their dramatic words.

Nonetheless, the fact remains that the story rings true to the Viking way of life. The Vikings did fight, sack, plunder, and settle in Europe and Britain during that time. They did behave as Ragnar and his family behaved. They shared the same world view. They lived according to the same set of values. That much is true, as historical records show and the people who came to hear the stories revealed by their pleasure in hearing them.

Ragnar and his sons exemplified the Viking way of life. Moreover, they did it all in the chaotic, turbulent, and often brutal world of emerging nations. Did Ragnar Lothbrok really exist? Almost certainly, the answer is yes, at least in some form.

Chapter 12

Ragnar Lothbrok: Hero or Villain?

According to the standards of Norse mythology, Ragnar was an utter failure, because he died a death that would have landed him in an icy hell. Yet, by the pre-Christian heathen religion from which he emerged, he did just as he should, following the predetermined path set out for him as all people do.

To the modern mind, this may seem unfair. After all, we want to believe we have control of our lives, and we think our heroes should have it, too. However, to understand Scandinavian lore, one must go beyond today's perspective and think like the people of that time thought.

Early Medieval times were so different from current times that you need to adopt a different mindset to understand them. The goals of human survival were different than they are now. Their resources were limited; they had to fight just to stay alive and thrive in the inhospitable world they knew at that time. Yet, their region was more overpopulated than other regions, making it advantageous for them to expand into other territories. In addition, they wanted merchant goods to meet their basic survival needs as well as their desires. Since Christians would not trade with them, the only way to get them was often to fight for them.

The Vikings relied upon their own set of values. For example, they would never attack an enemy that was already in battle. Ragnar Lothbrok and other Viking heroes also exhibited great courage in the face of danger – so much so that they went outside the bounds of their relatively safe homes to seek it out. Today, we celebrate those that explore new regions against incredible odds, so even to the modern mind, it makes sense that the Old Norse poems would glorify their exploits.

On the other hand, Lothbrok was so bent on competing with his sons that he embarked on many of his raids just to show them up. This is not the behavior we expect or condone from parents today.

Lothbrok and his sons seemed particularly ruthless and violent in their treatment of enemies, prisoners, and even townspeople in the areas where they invaded. In this sense, Ragnar can be seen as a villain. Still, these acts were no worse than those committed during holy wars throughout history, including the Christian Crusades.

Yet, the fact that such brutality has happened throughout history does not make it right or just. It does not matter that this was their way of dealing with hardship, especially if there were other means at their disposal.

In the time since the Ragnar tales, the descendants of Vikings like Ragnar have had a personal stake in the story. Each country celebrates those that paved the way for their success as a nation, just as many modern Danes salute the deeds of Ragnar and his clan. Ragnar was an invader, certainly, but he was also a Scandinavian liberator. He created new settlements throughout England and Europe, and his people often remained behind, pledging their allegiance to the kings of the lands they had invaded.

The great question each person must decide in choosing whether to hold up Ragnar Lothbrok as a hero or condemn him as a villain really has to do with whether one sees values as absolute ideals or as a function of the time and place in which they are espoused. When one considers these options, the answer becomes more apparent.

Ragnar Lothbrok might be thought a villain to those that lived in the areas he attacked or modern-day critics. However, as a loving husband, a courageous warrior and a generous commander in Medieval times, Ragnar Lothbrok can be considered a hero as well. Each person must make his or her own evaluation of Ragnar's character based on his or her own particular perspective and values.

Chapter 13

Ragnar's Story: Early Literary Works

Most of the information we have on Ragnar Lothbrok comes from literary sources rather than strictly historical documents. If the Vikings kept detailed records of their conquests, those documents do not survive today. However, storytellers called Skalds composed and memorized poems and stories to tell the tales of Viking history, legend, and mythology.

Skalds traveled from town to town, bringing news and education to Viking children as well as townspeople who had little other means to hear what was going on in the world. Children learned the thoughts and behaviors deemed acceptable by the Viking culture from these roaming storytellers.

The Skalds were not only educators of a sort, but they were also valued, entertainers. Many accompanied their tales with music played on harps or lutes. The stories they told were often reverent hero's quests. However, their tales and poems could also be funny, sarcastic, bawdy, or defiant. Because winters in the Scandinavian regions were harsh, people were happy to listen to the Skalds' amusing tales, laughing at their heroes' mischief or greed or drawing inspiration from their courage.

Many of the stories have been collected and written down, but this typically happened hundreds of years after the time the stories took place. Following is a list of literary and historical writings that recount the stories of Ragnar. Passed down through the oral tradition and molded into literary works or copied by scribes, these stories are sometimes inconsistent with each other. However, many of the tales are repeated in more than one of the different works.

The Saga of Ragnar Lothbrok

This Icelandic literary work was written down in the 13th century. Ragnar Lothbrok is the focus of the entire saga. Ragnar's saga is contained in a document that begins with the Volsunga Saga, which is the story of two hundred years of Viking history that includes stories about Aslaug's parents Sigurd and Brynhildr. One of the sources is Adam of Bremen, a Christian historian who lived sometime shortly after 1000 CE. Another source is Saxo Grammaticus' Gesta Danorum. The Saga of Ragnar Lothbrok continues with that story to cover Ragnar's life.

The Saga of Ragnar Lothbrok covers the lives of Aslaug and Thora Town-Hart. It also tells about the battles in which his sons fought. Finally, it relates the story of how King Aella killed Lothbrok.

Tale of Ragnar's Sons
This short story was written down hundreds of years after the events happened. It was probably composed in a monastery as a year-by-year chronicle of known historical events.

The Tale of Ragnar's Sons includes stories of Ragnar, his exploits and his death. However, it focuses on the tale of how the sons took vengeance on England for Ragnar's demise. The rest of the work tells stories about other Danish kings, as well as English and Norwegian kings.

Anglo-Saxon Chronicle
The earliest copy of the Anglo-Saxon Chronicle that still exists today was written down starting in

the 700's or 800's. Its sources include oral sagas and direct quotations told around 755 CE, indicating that it was at or near the time of the recording. It was published around 892 after the Vikings returned to England. Since it was written in England by Christians, it favors that country to the disadvantage of the Vikings and Scandinavia in general. Ragnar Lothbrok is not explicitly mentioned in this listing of events that go as far back as 60 BC.

Heimskringla Saga
Poet and historian Snorre Sturlsun wrote this saga about the Old Norse kings in Iceland. It seems to have been written around 1230 CE. The name of the saga means "The circle of the world." Although 19th-century historians accepted Sturlsun's depiction of Viking kings and their deeds, 20th century critics are less likely to agree with its accuracy. However, the sagas do seem to tell important information about Norse society and politics.

Saxo Grammaticus' Gesta Danorum ("History of the Danes")
This far-reaching history of Denmark was written by a Danish historian who lived between 1150 and 1220 CE. Although the history is mostly a product of Saxo Grammaticus' imagination, it

explains the values of the early Danes very well. This literary work includes stories of Vikings such as Ragnar Lothbrok, and Ragnar is mentioned by name.

Krakumal

Written down during the 1100s, Krakumal is the story of Ragnar Lothbrok's death. According to legend, Ragnar composed it while he was being killed and recited it before he died.

Fragmentary Annals of Ireland

This Middle English chronicle/history mentions Ragnall, son of Alpdan. Some proponents of a historical Ragnar Lothbrok cite this work as a clear record of his existence. The text is part history, part romance, but reading a modern translation of the surviving parts of the annals can give a different perspective on the Ragnar legend.

Prose Edda

This work of Snorre Sturlson refers to Ragnar's deeds and also is a relatively complete description of the Skaldic tales revolving around Norse mythology and Viking conquests. It is the basis of the Tale of Ragnar's Son and the Sagas of Ragnar Lothbrok.

Poetic Edda

Another collection of Old Norse mythology and legend is the Poetic Edda. It contains the most well-known of the anonymous poems of the Nordic people and the history of Iceland.

Conclusion

From the Early Middle Ages to modern times, the dramatic story of Ragnar Lothbrok has been told countless times. Skalds of Ragnar's time told the tales through the oral tradition, scribes wrote them down later, and people have been relating the story ever since. During the last several decades, the basic story has been included frequently in fiction books, movies, and television series to fascinate fans of this Danish hero. New audiences have been introduced to Ragnar and are often interested in finding out more about Ragnar's life and exploits.

According to legend, Ragnar Lothbrok was a lover, a fighter, a king, a commander, and a hero (or a villain). His story has become even more compelling as it has been dramatized and exaggerated over time, lending an almost mythic quality to events that had their origins in the reality of time, place, and person. This book has provided a brief introduction to those times, places, and people surrounding Ragnar Lothbrok, for the only way to understand legends is to ground them in the times and places in which they flourished.

Made in United States
Troutdale, OR
09/17/2024

22919800R00040